Introduction

Another 'fallen flag' is the name David Brown as associated with tractors. But there are some tractors which stand out above the rest in the history of mechanised agriculture. And David Brown is one of these names. The Company came into the tractor business almost by accident with a somewhat difficult partnership with Harry Ferguson, and sadly over the years had to play third fiddle to the big boys Fordson and Massey Ferguson.

The products from Meltham were, however, much thought of by farmers not only in the United Kingdom but throughout the world, and the revision of our original book to take the story up to the phasing out of the last true 'David Browns' makes this title of appeal to all who knew and used the marque, or simply those old and young who still restore and enjoy such tractors.

ACKNOWLEDGEMENTS

The author would like to thank the following for their assistance with information and illustrations for this book.
David Bate, Charles Cawood, Selwyn Houghton, John Melloy, whose assistance with the original David Brown book was invaluable, Bob Moorhouse, Stephen Moate, The Institute of Agicultural History at Reading University and all those customers who continue to send in their thoughts and information on all matters regarding David Brown and other tractors.

ISBN 1 85638 011 4

Design Layout and Editing by Allan T. Condie
Typesetting and reproduction by Rivers Media Services.
Printed in England.

© Allan T. Condie Publications 1996   All rights reserved.   Reprinted 2002
No part of this title may be reproduced in any form by any means, including photocopying, except for normal review purposes without the written consent of the Publisher. It shall not be disposed of at any other cover price than that shown thereon, or in any other cover or binding other than it was first supplied. All trademarks, model names and designations are used for identification purposes only.

Allan T. Condie Publications, Tel./Fax. 01455-290389.

THE DAVID BROWN TRACTOR STORY

The David Brown industrial concern can be traced back to 1860 when it employed only two people manufacturing wooden gear patterns for the textile mills in Huddersfield. By 1910, after steady expansion, the firm had become the largest manufacturers for gears in the British Commonwealth and by the end of the First World War a steel foundry had been incorporated.

In 1921 the second David Brown started work in the Huddersfield factory and worked up through the position of foreman, assistant works manager, works manager, director and was appointed managing director in 1932. David Brown was himself a farmer and during a business trip to America in the mid 1930s he was impressed by the large way in which American farming was being mechanised. He also noted the majority of tractors working at home in Britain at that time were American built. Using his own capital he rented space from the gear works and began to develop his own ideas on an agricultural tractor.

In 1929 another tractor pioneer, Harry Ferguson, had decided to build a tractor incorporating his linkage patent and by 1933 the "Black Tractor" was built and Ferguson began to look around for someone to manufacture this machine. David Brown agreed to build it, with its revolutionary features, and Harry Ferguson set up a new company to market the Ferguson-Brown tractor. By May 1936 the first tractor had been built at the Park Gear Works and was priced at 224. This tractor had a 4-cylinder water cooled petrol or petrol/TVO engine built by Coventry Climax - their type "E" unit, giving 20bhp at 1400rpm, bore 3 1/8", stroke 4", capacity 2010cc. Advertised as the world's first production tractor with hydraulic lift and converging three point linkage, sales were very slow to begin with, so Ferguson and Brown merged the manufacturing (DB Tractors) and sales company (Harry Ferguson Ltd) to form Ferguson-Brown Tractors Limited, with both men as joint managing directors.

This had no marked improvement in sales and David Brown was sure the tractor was in need of major changes in specification, but Harry Ferguson would not hear of any changes to his original design. There were various design and production problems, indeed in order to maintain continuity of engine supply David Browns built the type 'E' engine under licence for later tractors.

Harry Ferguson took himself off to the USA to show a Ferguson-Brown, and implements, to Henry Ford. This meeting produced the famous handshake agreement between the two men, which resulted in Ford manufacturing the new Ford-Ferguson tractor for Ferguson and, by June 1939, the first machines were ready for marketing.

The VAK1

This left David Brown on his own in Britain to go ahead with his ideas for a tractor but not before he had undertaken a personal round-Britain market research trip to find out from the farmers themselves what type of tractor they required.

The information collected was put to good use and an engine designed by A. Kersey was built (incorporating improvements suggested by Harry Ricardo for possible conversion to diesel) by the end of 1938.

The new tractor was designated VAK1 (vehicle, agricultural, kerosene, Model 1) and was shown to the public for the first time at the Royal Show in 1939 where its clean, functional lines made it an immediate success with the farmers and orders for 3,000 units were taken. Painted in the distinctive "hunting pink" the tractor developed 35hp at 2,000rpm from its 4-cylinder water coded petrol or petrol/TVO engine with 4-forward and 1-reverse gears, implement depth control was by depth wheel system when used with the power lift which was available along with PTO shaft and independent hand brakes, adjustable dished wheel centres gave track adjustment.

This large order for the tractor now gave David Brown scope to expand his business in the old United Thread Mill factory at Meltham Mills, but World War Two intervened, and the tractor production was severely cut (instead of producing 3,000 tractors in the first year, only 1,000 had been produced) in order to produce gearing of various types.

By the end of the war, the company had built 5,300 VAK1s and 2,400 VIG1s. This was the industrial version of the VAK1, used by the RAF to tow aircraft and bomb carriers, plus 110 DB4 crawler tractors, which were used by army engineers.

VAK1a and VAK1c "Cropmaster"

Now that the war was over the switch over to peacetime production of agricultural tractors was begun with the introduction of the VAK1a, in March 1945. This was a new development of the VAK1 tractor, but featured an improved governor and engine lubrication system, automatic load controlled manifold hot spot, plus the David Brown patented, turnbuckle top link.

These early built Meltham machines earned the company a reputation for reliability in the highly competitive tractor industry and in April 1947 the first "Cropmaster" (VAK1c) model was introduced. This was a development of the former VAK1a and with its up-to-date specification it must have been one of the most comprehensively equipped machines available at that time. With its introduction, the Cropmaster boasted as standard many items which until that time had been regarded as extras, for example hydraulic lift, swinging drawbar and electric lighting. Transmission was through a Borg & Beck single dry plate clutch, and 4 or 6 speed gearbox to the differential and final reduction units. Clutch was independently controlled by pedal or hand lever. During its production run, the Cropmaster was developed further, and a diesel model was introduced in November 1949. Super, Prairie and Vineyard all followed during the long production run enjoyed by this tractor, from 1947 to early 1953.

Another development of the Cropmaster was the Trackmaster crawlers, available from 1950 (in either petrol/TVO or diesel form), the Industrial Cropmaster M and the Taskmaster. A brief description of the above mentioned machines are as follows:

Vineyard: A narrow version of the Cropmaster, introduced in 1947, 179 being made until withdrawn in March 1953. A diesel-engined version was introduced in 1951, but only about 80 were built in nearly 2 years.

Super: Introduced in October 1950, the Super Cropmaster used a bigger engine, large section tyres, full engine side panels, chaff screens to the grille and a rain trap in the exhaust silencer. The production run of this model lasted until December 1952 in which time 4,880 units had been built.

Prairie: This was a high clearance version, available in October 1951, until April 1954, also made with diesel engine from November 1952, withdrawn in April 1954. Total units produced were 763 (519 plus 244 diesel).

Cropmaster M: Tractor, less hydraulic lift, from April 1947 until November 1954. Also MD (diesel from July 1950-January 1956).

Taskmaster: Similar to wartime VAK1, from August 1948 to August 1953. Also in diesel form from May 1951 to September 1953.

Trackmaster: Crawler version of the Cropmaster, having the same engine but a new 6-speed transmission, incorporating differential steering. Built January 1950 until March 1953 in petrol/TVO form, of which 700 were manufactured. The diesel version was designated the Trackmaster Diesel 30, available from September 1950 until January 1953, and approx. 500 were made.

In April 1952, DB had introduced a completely new crawler called the Trackmaster Diesel 50, which use a 6-cylinder, 50hp diesel engine, a 6-speed gearbox and differential steering. These large machines were made under the Trackmaster banner until April 1952, when about 150 had been produced. December 1952 saw the way going towards the Super Cropmaster. In January

The 50D

1953 the new 50D wheeled tractor made its first appearance. This machine used a 6-cylinder 50hp engine from the crawler and was intended for towing operations and belt work, having a side mounted belt pulley, 4-speed TVO, and was available only with a diesel engine. Only 1,260 of these tractors were manufactured in a five-year production run, and the like was dropped in June of 1958.

The 25 and 30 Series

February 1953 saw the introduction of the new DB 25 Series tractor, available firstly in petrol/TVO form, followed in October of that year by the 25D. Both tractors had 3.5" bore X 4" stroke engines, developing 31bhp at 1800rpm, and proved to be popular small tractors until discontinued in 1958.

These tractors were easily identified by the lack of tinwork - usually associated with DB tractors - namely the famous horseshoe scuttle, full width wings and the disappearance of the bench seat, although some very early units retained these features.

March and April of 1953 saw the introduction of the DB 30C and 30D tractors, which retained the Cropmaster Series' styling, until replaced by the new Series tractors in April 1954. These new tractors had the same tinwork as the 25 Series but were available from new with TCU (traction control unit). This was a controlled weight transfer system which allowed smaller tractors to carry out heavy work using larger implements. The unit was upgraded in 1955 to allow the hauling of heavy loads using a special trailer hitch.

The 2D

Introduced for precision market garden type work and rowcrop use, the 2D offered a 2 cylinder diesel engine mounted on the rear of the frame, to allow the use of mid-mounted implements. The lifts (rear lift and PTO were optional extras) were operated by compressed air and the front cylinders could be used independently. The 2D was introduced in March 1956, followed by a narrow version, in July 1957. Both models were discontinued in September 1961.

Crawler Update

From 1953 the crawler range consisted of the following:

30TD, 30T and 50TD, which were agricultural machines, 301TD, 501TD Series took care of the industrial side of the market, in 1957 a MkII 50TD was introduced, using a larger diameter clutch and new running gear.

The other machines were superseded by a new 40TD model in 1960, which used a 40hp diesel engine. This crawler was produced for three years and was finally discontinued along with the 50TD in January 1963. Other machines built at this time were 301C, 301C to Air Ministry specification, 301D, Taskmaster Turbo Petrol and Diesel models and Medium-wheeled Aircraft Towing tractors.

Albion

1955 saw another important addition to the company with the acquisition of the Harrison, McGregor & Guest Implement Manufacturing works at Leigh, with its famous Albion range of agricultural machinery. This gave DB a much wider range of implements, consisting of ploughs, ridgers, rotary tillers, mole drainers, muck spreaders, seed drills, mowers, swath turners, pick-up balers, forage harvesters, binders, combines and front loaders by the late 1950s, to upgrade the implements introduced with the early VAK tractors, before the war.

900 to 950

Available with four types of engine, the new DB 900 introduced in November 1956 offered 40hp from the diesel, 37hp from the TVO, 40hp from the petrol and 45hp from the high compression petrol model. This new tractor was painted red with blue wheels and used, for the fist time, the distributor type injection pump. Other points of note were the removable one-piece bonnet and dual category linkage. The tractor range was further improved the next year with the introduction of the 900 live drive, with dual clutch giving live hydraulics and PTO.

By November 1958, the 900 was replaced by the first of the red and yellow 950 TU Series tractors, which used a 42hp diesel or petrol engine. Also, improved steering and drawbar. Further improvements were introduced to the 950 range with the 950 Implematic V and W series, which were superseded by the A and B series tractors, in 1961.

The 850

The "Implematic" name also appeared on the 850 4-cylinder tractors, introduced in 1960, which gave farmers the opportunity to use depth wheel or draught control. These early 850 models were also designated A and B series, and produced about 35bhp in diesel or petrol form.

The later C and D Series only had diesel engines, plus multi-speed PTO and improved front axle clearance. This tractor was also available in narrow form.

990 and 880

By 1961 the DB tractor range was further improved by the 990 and 880 "Implematic" tractors; the 990 went on to be the best selling Implematic tractor produced, and accounted for nearly 50% of production. Power was from a 52bhp direct injection diesel engine, with 3 5/8" bore and 4 1/2" stroke. During its production run such improvements as height, control, increased wheel base, offer of 12-speed transmission, and the fitting of the battery up front, along with the two stage air-cleaner, were all introduced. The smaller 880 Implematic tractor used the same engine as the 950 model and was described as "the ideal tractor for the smaller farm", offering a higher speed range than the 950, from the same 42.5bhp (diesel only) unit.

By 1964 a new 880, with 3-cylinder diesel engine, was on offer, and the old 880 and 950 were discontinued.

Enter the 770, 880 and 990 Selectamatics

Early in 1965, DB announced a new, small tractor, in the shape of the 770 Selectamatic. This was powered by a 3-cylinder diesel engine, giving 33bhp and used the Selectamatic's hydraulic system and 2-lever 12 speed gearbox.

October 1965 saw the introduction of a whole new range of models in the colour scheme of orchid white, and metallic chocolate brown, to mark the introduction of the 990, 880 and 770 Selectamatic tractors.

The 770 was the first tractor to be fitted with the Selectamatic hydraulic system, powered by a 33bhp 3-cylinder engine. This tractor was upgraded in October 1965 to 37bhp, and used a 2-lever, 12-speed gearbox as standard equipment.

Due to the success of the Selectamatic hydraulic system, used on the original 770, the new 55hp 990 and 46hp 800 were fitted with this new system, along with multi-speed PTO and differential lock. Three forward, 4 reverse speed gearbox was an alternative, along with high clearance conversion.

By 1967, the 60+ hp tractors available in the UK included the DB1200, which offered a hand clutch for the drive to the PTO. The 3 point linkage was available only for Category 2 attachments and the hydraulic pump was driven direct from the front of the engine.

Also, in 1967, the 780 tractor was introduced as a live drive model, with a two stage clutch. 1968 saw the 1200 uprated to 72hp and safety cabs were available for all models.

Announced in July 1970, the 1200 and 990 four-wheel-drive tractors were available (the 900 in 12-speed form only).

In July 1971 the new Meltham tractor assembly complex was in operation and by August of that year the 1212 Hydra-shift, with semi-automatic transmission was in production nearly 10 years on from the ill-fated 990 Auto-drive Implement model. The 1212 Hydra-shift was fitted with power steering as standard.

The other tractors available by this time were the 990 Selectamatic, 990 Synchromesh, 990 Selectamatic 12-speed, 995 Synchromesh, 1200 Selectamatic, 995 Synchromesh 12-speed, 1210 Synchromesh and 1210 Synchromesh 4-wheel drive.

The last tractors introduced to truly David Brown designs were the 1410 and 1412 which were the first DB tractors to feature turbocharged engines.

David Brown Tractors Limited were acquired in 1972 by Tenneco, an American conglomerate, which already owned Case, and by 1973 David Brown and Case tractors were appearing in the new colours of orchid white, powder red, and black. In some territories the tractors were devoid of any David Brown badging and were sold as Case. Conversely, Case models were sold with David Brown badging on the UK market.

Opposite top: Type 'A' tractors being assembled at David Brown's Park Works in Huddersfield. Production was on the flow line principle. This tractor had its shortcomings, but it brought David Brown into the agricultural business.

Below: An aerial view of the Meltham Mills complex.

The type 'A' is rightly covered in more detail in our companion volume 'The Ferguson Album'. Nevertheless, as a David Brown built machine, we illustrate it here. Opposite page (from top to bottom), shows the tractor with its two furrow plough, hauling a horse rake, and in action with cultivator. The hydraulic lift linkage can be seen above, whilst a complete tractor on pneumatic tyres can be seen below. Certain components on this tractor, such as wheel bearings, continued in use on later D B models.

Opposite page: above: VAK1 tractor with steel wheel equipment. Earlier tractors had a different type of heat indicator fitted.

Below: Mr David Brown, on left, shows the new tractor to world land speed record holder, Captain G.E.T.Eyston. Note the early type wheels, cast iron grille, and proper silencer equipment.

Above: The power lift unit for the VAK1 was not fitted as standard and many tractors would leave the factory like the one shown here at work with a Ransomes trailed plough. D B made their own trailed plough between 1939 and 1944 for tractors purchased without a lift.

Below: Implement depth was controlled by a patented depth (gauge) wheel system. Note the hand levers for operating the independent brakes.

Above: Introduced in March 1945, the new VAK1a tractor incorporated a number of improvements over the previous model, namely, to the engine lubrication system, better governor and an automatic 'hot spot' for rapid engine warm up. A new grille was used with horizontal brass David Brown badge mounted on the front of the bonnet.

Left: The VAK1a could also be had with steel wheel equipment, 16 track widths could be obtained by moving the dished wheel centres and reversing the rims.

Below: Wartime tractor production was mostly for the Royal Air Force and Navy (Fleet Air Arm), for heavy towing and winching duties. Here, we see a VIG/100 adapted for threshing work with the forward mounted pulley unit. These tractors were commonly known as 'Thresherman'.

The 1947 model, designated VAK1c, was universally known as the Cropmaster. New features included, the adoption of a two piece frame, improved final drive, a six speed gearbox option, dual application of the steering brakes by foot pedal, a two speed PTO, A wider seat, and the offsetting of the steering wheel 2" to the right to give more driving room. The nearside of the tractor is seen below, whilst the offside is seen above in this ploughing view.

Above: Close-up of the older type Cropmaster, showing the grille and front end. This tractor would appear to be fitted with 'Field Lighting'.
Below: A later Cropmaster, with altered front casting and underslung weights.

Above: A line-up of Cropmaster tractors at a field working demonstration in Coupar Angus, Perthshire, by Grassicks Garage Ltd. The date was 18th March 1948 and the location, Easter Denhead Farm.

Below: Visitors to the Meltham complex get their first opportunity to try out a Cropmaster. This view shows a TVO model in the foreground and a diesel behind.

Above: November 1949 saw the introduction of the famous Cropmaster diesel tractor; The dimensions of the engine were the same as the TVO engine, and conversion kits could be obtained.

Below: For export territories, the Cropmaster was equipped with vee twin front wheels. Note the battery carrier, heavy duty steering column and large steering wheel on this diesel example.

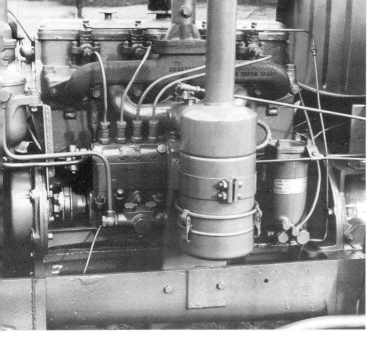

Left: Nearside view of the Cropmaster diesel engine. David Brown was the first major manufacturer to introduce an 'own-make' diesel engine onto the UK tractor market. It was an engine which was to continue in use for some years.

Below: A number of sectional models were built for use at shows. These were usually built up by apprentices using as many components which were deemed as 'scrap' insofar as normal production was concerned.

Bottom: Early type Cropmaster with cab and 9" rear tyres and rims.

Above, and below: A vineyard version of the Cropmaster was produced from 1947-53. This featured a reduced width front axle and the rear wings set in to enable the rear track to be reduced.

Above: The Prairie Cropmaster was produced from October 1951 to April 1954 and was specifically aimed at the export market. Higher clearance was obtained by fitting oversize tyres. 12-volt electrics were standard, and this model was the first to eliminate the 'scuttle' in front of the driver and to use a single pan seat as standard.

Above and left: The Super Cropmaster was introduced in October 1950 and featured an uprated engine, running at 2300rpm, which gave 28bhp. Other features were, the 13x28" rear tyres, full bonnet side panels, the abandonment of a pre-heater in favour of a large mushroom shaped pre-cleaner, and 12 volt electrics as standard, with starting and dual headlamps. This latter feature, plus the chaff screens on the grille, readily identified the model, which was offered until December 1952. (Normal VO Cropmasters had 6-volt electrics, except when 12 volts was specified for export).

Above and below: The 50D was launched in 1953 and used the same 6-cylinder engine as the 50TD Crawler. It was basically designed for towing operations and therefore no provision was made to fit hydraulics. In the five years of production only 1260 units were completed. The early version is seen above, whilst the later production shows some important changes such as the location of the air cleaner and headlamps.

Above: In design the 50D was very much an overgrown 'Cropmaster' with many family features. The engine was a unit of 35/8" bore by 4" stroke, using common components with the 30D A six speed transmission was provided and the rating was 50bhp, or five furrows. Unusually for David Brown practice, a side belt pulley was fitted. Indeed, earlier tractors had a frame mounted aircleaner and a smaller side pulley; the aircleaner had to be moved as it fouled the belt position. The engine, driving position, PTO and drive pulley are seen in the shots below. Note the similarity of this tractor to the Nuffield 'Universal'. The reason for this is that Dr. Merritt, who had been working at David Brown's during the War, had done much preliminary work on the postwar range, but then left to take up a position with the Nuffield Organisation.

During the Second World War, production of tractors was severely restricted - David Brown being obliged to concentrate on other work - but from 1941-44, the VIG1 industrial model was supplied to the RAF, later units having fluid flywheel transmission. A winch was fitted for recovery work. The two views here, show the VIG/100 tractor to advantage, the full mudguards being a feature of most David Brown industrials form that time on.

Above: A VIG/1 with front guard, rear wheel weights, footboard weights and rear towing light.
Below: A standard model VIG/1.

Above: A becabbed version is seen in 1950, in the employ of a Hebden Bridge contractor.

Below: The postwar designation for industrial models was 'Taskmaster', and the view shows a VIG/1 model in 1948, hauling a cattle box.

Above: A number of VIG tractors were sold new, and indeed a number of ex-RAF machines also, to various civil airports up and down the country. Here is one at work, hauling an aircraft, the job for which it was designed.

Below: The VIG was available in various forms, this is a mid 'fifties example with dual rear wheels, front guard, extra weights on the footboards, and special wide wings on the rear.

In 1953, the Cropmaster was superseded by the 30 and 25 series of tractors. A 30C is shown above, in later form - early examples retain the Cropmaster mudguards, scuttle and double seat. The engine of the 30C was of 3.625" bore by 4" stroke and gave 37.6bhp at 2300rpm. Other new features were the wings and pan seat, and the adoption from 1954 of TCU (traction control unit). The shot, below, shows a 25D with JCB halftracks.

The above view shows the 30D with its 34bhp engine. The two six-volt starting batteries can clearly be seen. Note also the pulley and PTO guards, fitted some 11 years before this was mandatory.

Below: This DB25 is seen carting grain from an early class combine on the NCB farms, in East Lothian, Scotland. Close inspection of the photo shows that the combine is powered by a D B unit.

Above: A 25D, ploughing on the company's own farm in Yorkshire. This was used for taking numerous publicity shots.

Below: The 25C could readily be identified by the lack of large batteries and the air intake coming from under the bonnet. 4.00 x 19" front tyres were fitted as standard.

The 25D tractor, seen from all angles in these three views.

Left: This rear view shows clearly the three point linkage, with adjustable top link and rear drive pulley attachment.

By the mid-fifties, the David Brown 'image' was getting a bit dated and the horsepower range was somewhat low, with the trend towards bigger implements. Things were to change, however ......

Above: The David Brown stand at Smithfield in December 1954?
Below: Exports were important in the 1950s and a show stand in Italy is seen here.

Three views of the 2D tractor, which was a 1956 introduction to the range. Its lightweight, air cooled, rear mounted, 2 cylinder diesel engine of 3+" bore, by 4" stroke, gave 14bhp at 1800rpm. It was basically designed as a market garden unit but could be used for rowcrop work on large farms. The front mounted implements could be lifted by an air assisted lift. The tractor was produced for 5 years until 1961 and some 2,000 were made.

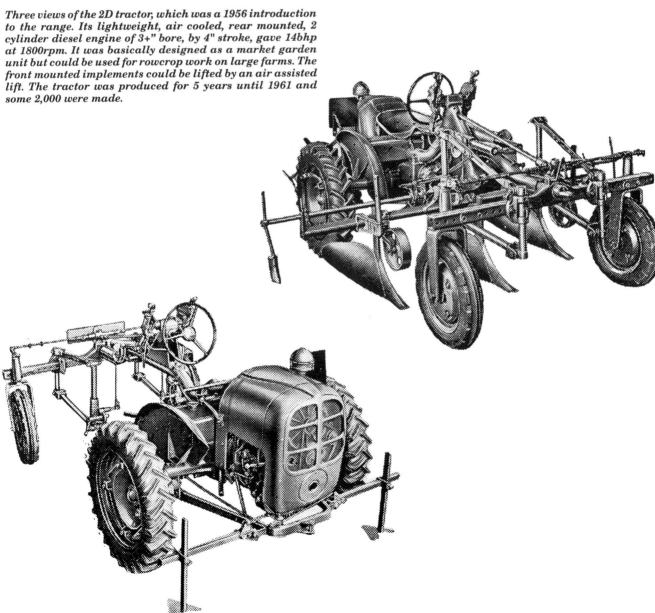

Opposite page: In 1956, the 900 was introduced, marking the first shift away from the total 'hunting pink' colour scheme, featuring blue wheels and grille inserts. It was available with a choice of engines, all based on the units previously used in the 30 and 30D, with a 35/8" bore and 4" stroke, with option of diesel, VO, petrol and special high compression petrol. A six speed transmission was employed and, from 1957, a live PTO and hydraulics were introduced. Both views here show the diesel model and the distributor type fuel injection pump is clearly seen. The traction control unit was a feature of the hydraulics. A high clearance model is shown (lower), with single front wheel. The Rowcrop Tricycle D B 900 tractor was fitted with a 9.00 x 10" front wheel and 11.00 x 39" rears. Optional equipment for this model included, dual-control valves for front and rear mounted toolbar use, centre mounting brackets, centre tool frame lift, lift ram and depth control stop, 10' centre-mounted monoframe toolbar, and 8' rear-mounted monoframe toolbar. A number of other features on all 900 models were new including, the detachable bonnet, an adjustable front axle, where applicable, and the diesel engine which developed 40bhp at 2000rpm. Over 13,000 were made.

This page: Three interesting views of agricultural show stands in the mid 1950s with a wealth of equipment to inspect.

Above: The 900 developed into the 950, which was produced from 1958-59, and this tractor gave increased power from the engine - through design improvements - the diesel now giving 42.5bhp. A new type steering unit was fitted to this model and an improved drawbar to BS specification, fitted. In addition, the colour scheme was amended to give cream grille, wheels and exhaust, and it became more usual to see this machine on 11 x 32" tyres, the 900 usually had 28" rears.

Right: This is the 'Livedrive' version, undertaking cultivation work.

Below: Between 1960 and 1963, DB manufactured 2,148 tractors for the Oliver Corporation of America. The tractors had a restyled bonnet, green and white paintwork, and were known as the Oliver (DB950) and 500 (DB850).

The 950 is seen from above, and provides an interesting comparison with the similar view of a 30D shown previously. The 950 'Implematic' superseded the previous model in 1959 and ran until 1962. Draft control, once protected by Ferguson patents, now expired, and were available, for the first time. In 1961, a new design of front axle was fitted to improve clearance. In addition, a handbrake was now fitted, for parking. Lighting was standard equipment. A differential lock was fitted on this model.

Right: The 4-cylinder engine is shown, by now a well tried design, dating back to the original Cropmaster diesel, of 1949. Now developing 52.5bhp, this unit was given prominence in sales literature, as demand for TVO and petrol units was waning. These latter were still available, if asked for, however.

Above: David Brown returned to the smaller tractor when the 850 was introduced in 1960. This unit used many parts in common with the 950 but, by returning to the original 3+ bore of engine and an engine speed of 2,000rpm, 25bhp was given. 28" rear tyres were fitted as standard, and the hydraulics gave all the features of the larger model. The A and B series, produced from 1960-63, were also available with petrol engines but, from the introduction of the C and D series, diesel engines only were available and multi speed PTO fitted as standard. The illustration shows the left side of the tractor. Note the aircleaner is still mounted in that time-honoured position. From 1963, height control was fitted to the hydraulics. This was the 1960-63 Oliver 500, as sold in the USA.

Below: In 1961, the 990 'Implematic' was introduced, this featured a redesigned engine with cross flow cylinder head, developing 52bhp at 2200rpm. The 3.625" bore was retained, but a stroke of 4+" was adopted. Initially, the battery and aircleaner were located as the 950, but from 1963, as shown here, these were relocated behind the front grille in front of the radiator. In addition, a 12 speed transmission became available.

Left: The 990 was available in industrial form, equipped with extras to comply with the Road Traffic Act requirements for tractors used for general haulage on the public roads in addition to 'on site' capabilities.

Right: This is the 880, also in industrial guise, Finish was industrial yellow.

Below: The DB 990 was shown at the 1961 Smithfield Show as the world's first production tractor to be fully equipped with a fully automatic transmission' - this was the DB 990 'Autodrive' the gearbox controls are seen here.

Below right: On the Continent, concern was already being expressed about the number of overturning accidents involving fatalities. The Scandinavian countries were among the first to legislate, and the 990, is here equipped with a frame which could be filled in to provide additional protection.

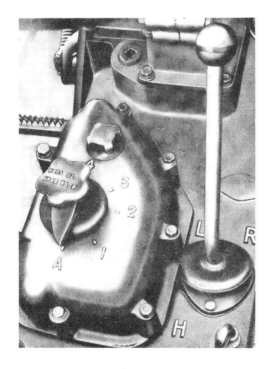

The 880 was the successor to the 850, in effect, although it was produced concurrently until 1965. It was basically the 850, with the 950s engine, but a three cylinder power unit was developed with the E and F series, with a 3 11/16" bore and 4+" stroke of engine, which developed 42.5bhp at 2200rpm. With the fitment of this unit, the oilbath aircleaner and battery were relocated in front of the radiator. The upper view shows the tractor with a four cylinder engine, while the lower shot shows a three cylinder model.

The 1965 introduction of the 770 Selectamatic brought in the Selectamatic hydraulic system, which was subsequently adopted for all other models. It featured a similar engine to the 880 with 3 cylinders but, in giving only 33bhp, it had an engine of only 4" stroke. The 12-speed gearbox was standard. The vertical injection pump is of note, a feature also seen on the 3 cylinder 880. There were not many 'hunting pink' 770s.

In October 1965, all DB tractors were restyled and painted in Chocolate brown and orchid white. The Selectamatic hydraulic system was now used on all models, along with multi-speed PTO and differential lock. The 990 was now rated at 55hp and the 880 at 46hp. The 990 was available with a 12-speed gearbox. An early example is seen above in standard guise, whilst the tractor below is a high clearance model. The East Yorkshire registration with 'C' suffix also dates the tractor at 1965. The opposite page shows later examples of the 990 after the Tenneco takeover, showing the ungainly safety cab originally fitted to David Brown tractors both 'dressed' and 'undressed'. This cab proved unpopular because of its overall height.

The 995 and 996 tractors featured a larger engine, the stroke being 4.5" rather than 4" on the 990. Whilst the 995 had a standard 'live' PTO arrangement the 996 had a fully independent PTO; however this feature was available as an option on the 990 also. Examples of the 995 and 996 with original style safety cabs are seen on the opposite page, whilst a 990 for export with seat fitted on the fender and safety frame is seen above. For export to certain territories the tractors were badged as CASE and with rollbar plus heavy duty air filter this tractor is destined for the Western Hemisphere.

Two illustrations of the 990 in export guise, again badged as a CASE. Note the power adjusted rear wheels on the lower example, power assisted steering. and the roll bar provided for certain export territories, complete with canopy in the top illustration.

Above: David Brown began to offer 4WD as a factory fitted option from 1970, in line with what other manufacturers were doing. This illustration also shows the DB 'Q' cab offered on all David Brown models from 1976.

Below: The 3 and 4 cylinder power units as used in 1965 David Brown tractors,

the four cylinder engine being originally used in the 990 and the three cylinder engine in the 770 and 880. A shorter stroke gave reduced power in the 770.

Right: The 880 as revamped in the chocolate and white colour scheme introduced in October 1965. The engine now gave 46HP.

Below: A later example showing different badging. The David Brown Selectamatic system offered 'live' hydraulics operated by a single control lever.

Above: The 885 Synchromesh was introduced in 1971 and is seen here without cab cladding. The gearbox gave 12 forward and 4 reverse speeds.

Below: An 885 with full cab equipment. These cabs were not popular due to their overall height restricting use in areas with low buildings.

Right: The 885 narrow was offered with 60" width for orchard applications. The DB live Selectamatic hydraulics gave four functions selected by a single lever - Depth (draught) control, Height (position control), TCU (controlled weight transfer) and external services.

Below: For export the 885 was equipped with a roll over bar and in this case it is badged as a CASE.

Above: An export 885 equipped with front end loader.

Below: The David Brown assembly line in the early 1970s with 995 tractors being finished off.

Opposite page: The smallest offering from David Brown was the 770 Selectamatic, introduced as a white tractor in October 1965, when the engine output was increased to 37HP.

March 1967 saw the introduction of the 67hp D B 1200 Selectamatic tractor. This model had a separate hand clutch, controlling the drive to the PTO, and the hydraulic pump was driven direct form the front of the engine. The engine was uprated to 72hp, in 1968.

Left: The 1210 in export form without cab. As with the smaller models a 12 speed gearbox also gave 4 reverse gears. The hydraulic pump was engine mounted.

Opposite page: Introduced in December 1971, the D B 1210 Synchromesh was initially offered in two wheel drive form. It is seen in original finish in the upper illustration and with orange engine after the takeover below.

Right: By April 1972, a 4-wheel-drive model was available.

Below: The 1212, seen here in export form with power adjusted rear wheels, offered the new 'Hydra-Shift' clutchless gear change which had four ratios within each of 3 forward and 1 reverse ranges. It was introduced in 1971.

Above: The 1212 for UK domestic sale with safety cab.

Right: The 1410 with four wheel drive and lower profile cab is seen here. This model was launched in 1976 and was the first DB tractor to feature a Turbocharged engine other than certain earlier industrial models.

Left: The 1410 for export is seen here. This tractor was launched to compete with the Ford 7000 and Massey Ferguson 1100 tractors.

Below: The 1412 was of course fitted with 'Hydra-Shift'.

Above and below: 'Q' cab legislation required all new tractors sold in the UK from 1st September 1977 to be so equipped, and such cabs had to be capable of keeping the noise level inside below 90db. This 1412 is equipped with the new David Brown luxury 'Q' cab.

Above: This 120 for export with four wheel drive did not legally require the fitting of a 'Q' cab - a rollover bar sufficed. Note the CASE badging for sale in the Western Hemisphere!

Below: A 1410, also for sale in North America. The 1400 series continued in production until 1980.

Above: The first D B Crawler was the 1942 DB4, which was basically produced for army use. It was fitted with a four cylinder Dorman Ricardo diesel, five speed gearbox, and clutch and brake steering. The engine was a 38.5hp unit, (above).

Below: The original Trackmaster, based on the Cropmaster. This example has an external air cleaner.

Above: Front view of an early Trackmaster, the early type of front casting is clearly visible. Note the deeper bonnet used.

Below: A view of the controls of the same machine, showing the early-type dash layout and hand levers.

Above: The basic Trackmaster 50. Fitted with a 50hp, 6-cylinder diesel, six speed gearbox, and differential steering, this model enjoyed a long run form 1952-65. Over the years, improvements were made, these included clutch modification and improvements to the tracks. The studio shot below shows side panels fitted.

Above: From 1950, the Trackmaster 30 was produced. Available with the same engines as for Cropmasters, the original Trackmaster designation gave way to the 30T and 30TD names, when the 25 and 30 wheeled models came out. This is a diesel example. Controlled differential steering was adopted on these units, with a six speed transmission.

Below: The 40TD was the successor to the 30TD and is seen above. It employed the improved engine with 35/8" bore and 4" stroke. Many later D B Crawlers were finished in a yellow livery.. The radiator guard and sump guard are to be noted.

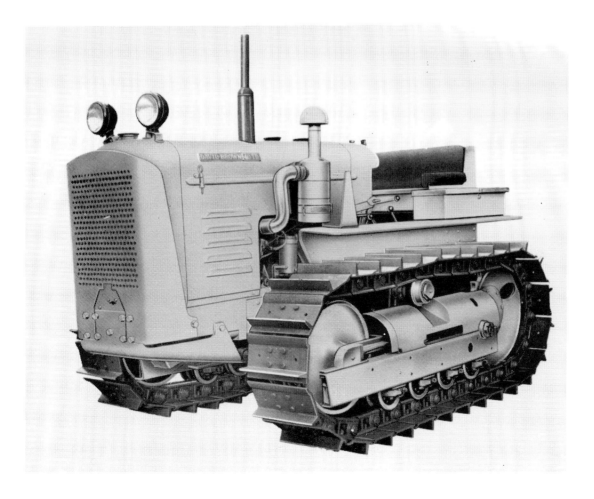

Above: A 50TD is seen here, at work. Note the blade and cab. No crawlers were produced by D B from 1965, but in 1972, after the Tenneco takeover, the association with Case's saw the use of D B engines in that concerns' crawlers.

Below: The 30 series, for industrial use, were designated 30IT and 30ITD, one of which can be seen here equipped with a dozer blade and hauling an Onions Empire' scraper.

DAVID BROWN TRACTORS
MODELS AND PRODUCTION DATES

TRACTOR DESIGNATIONS ARE BUILT UP AS FOLLOWS

PREFIX		PREFIX	
V	-Vehicle	9OOJ	
A	-Agricultural	950T	
I	-Industrial	95OV	
R	-Road Traffic Act Vehicle	950A	
C	-Cropmaster type	850A	
T	-Trackmaster	85OC	- Livedrive (Dual clutch)
K	-Kerosene	8SOC	-
G	-Gasoline	880E	-
D	-Diesel	880A	-
1	-First Series	990A	-
3	-3 518 in. bore engine	770A	-
6	-6-cylinder	90OL	-
p	-Prairie	95OU	
100	-With hydraulic lift	950W	-
200	-Without hydraulic lift	950B	-
U	-American model	850B	-
G	-German model	850D	- Standard model
		880D	- (Non-livedrive, single clutch)
		88OF	-
		880B	-

SUFFIX			
M	-Without hydraulics and PTO		
S	-12-speed gearbox	V	-Vineyard model
N	-Narrow model	H	-High clearance

ENGINE DESIGNATION BUILT UP AS FOLLOWS

-	Agricultural	4	4-cylinder
-	Kerosene	*3	3 5/8 in. bore
-	Gasoline	(when used at end of designation)	
-	Diesel	*3/3½	3½". bore
-	Industrial	6	6-cylinder
-	Stationary	2	2-cylinder
-	Marine		3-cylinder
-	Cropmaster and 3OC type		
(when used in front of engine designation)			
-	Lamp oil		
-	High altitude		

When not included in engine designation, denotes 3½". bore for 4-cylinder engines, excepting the IG/1C which has 3.625" bore. The 6-cylinder engine also has 3.75". bores.

From the 3-cylinder 880 and 770 Selectamatic engines onward, the final designation figures relate to the approximate capacity in cu.in. for 1 cylinder.

David Brown Model Designations & Production Dates.

Industrial Vehicles

Description	Class	Serial Nos.	Production.
Heavy Duty Industrial	VIG1/100	AW1-750	1940-1945
Heavy Duty Industrial	VIG1/100A	AWC1-	1945
Heavy Duty Turbo	VIG1/462	AW750-	1940s
Cropmaster M	VAG/1C/M; VAK/1C/M	M10001-10205	1947-1954
Cropmaster Diesel M	VAD/1C/M	MD10001-10012	1950-1956
Taskmaster	VIG/1AR	R10001-10392	1948-1953
Taskmaster Diesel	VID/1AR	RD10001-10144	1951-1953
(Taskmaster) 301C	VIG/1AR2	R 10325-10415	1954-1956
(Taskmaster) 301D	VID/1AR2	RD10001-10159	1954-1955
301C (Air Ministry)	VIG/1A/AM; VAG/1A/AM2	AM101-547)	1954-1956
301D (Air Ministry)	VID/1A/T/AM	AM101-547)	1954-1956
Medium Aircraft Towing (301C & 301D)	VIG/1A/T/R; VID/1A/T/R	HV101-413	1952-1956
Medium Industrial	VIG/1C		1953-
Industrial Towing	VIG/1C1		1956-
(900) Taskmaster Petrol	VIG/1H/R2	R10416-	1956-1959
(900) Taskmaster Diesel	VID/1H/R2	RD10160-	1956-1959
Medium Aircraft Towing (900)	VIG/1H/T/R; VID/1H/T/R	HV	1956-

Crawler Tractors

Description	Class	Serial Nos.	Production.
VAK on tracks	TAG1	c.550 produced	1940
DB4	CAT D4 copy	110 produced	1942-1949
DB5 experimental			
Trackmaster	TAK3; TAG3	10001-10700	1950-1953
Trackmaster Diesel	TAD3	10001-10500	1950-1953
(Trackmaster) 30T	TAK3; TAG3	10750-11153	1953-1958
(Trackmaster 30TD	TAD3	10550-11203	1953-1959
Trackmaster Diesel 50	TAD6	10001-10150	1952-1953
(Trackmaster 50TD	TAD6	10175-10686	1953-1956
30ITD	TID3	10001-10343	1953-1959
50ITD	TID6	10001-10298	1953-1956
50TD MkII	TID6	20001-20471	1957-1963
40TD	TID3	30001-30143	1960-1963

Threshing Tractors

Description	Class	Serial Nos.	Production.
Heavy Duty Threshing	VTK1	As VAK & VAK1A As VIG1/100 & VIG1/462	1940s

Agricultural Wheeled Tractors

Description	Class	Serial Nos.	Production
AK1	VAK1; VAG1	1001-6000	1939-1944
AK1A	VAK1A; VAG1A	6351-9852	1945-1947
ropmaster	VAK1C; VAG1C	P10000-44000	1947-1953
)C First Series	VAK1C; VAG1C	P44500-46058	1953-1954
)C Second Series	VAG1D/30; VAK1D/30	P/30 10001-12766	1954-1957
ropmaster Diesel	VAD1C	PD10001-18000	1949-1953
)D Diesel First Series	VAD1C	PD18500-20797	1953-1954
)D Second Series	VAD1D/30	PD/30 10001-19452	1954-1957
uper Cropmaster	VAK3/1C; VAG3/1C	SP10001-14881	1950-1952
rarie Cropmaster	VAK3/1C/P; VAG3/1C/P	SPP10001-10600	1951-1954
rarie Cropmaster Diesel	VAD3/1C/P	PDP10001-10245	1952-1954
ropmaster Vineyard	VAK/1C/N	N10026-10205	1947-1953
" Vineyard Diesel	VAD/1C/N	ND10001-10084	1951-1952
)D	VAD6	VAD6 10001-11260	1953-1958
5	VAG1C/25; VAK1C/25	P25 10001-21318	1953-1958
5D	VAD1C/25	PD25 10001-23424	1953-1958
)	VAD12	VAD12 10001-11644	1956-1961
)V	VAD12V	VAD12V 10001-10364	1957-1961
)0 Petrol	VAG1H/30	900G 10001-10305	1956-1958
)0 Kerosine	VAK1H/30	900K 10001-10265	1956-1958
)0 Diesel	VAD1H/30	900D 10001-17437	1956-1958
)0 - J & L Series	VAG1J/30; VAG1L/30 VAK1J/30; VAG1L/30 VAD1J/30; VAD1L/30	J900G; L900G J900K; L900K J900D; L900D 50001-55445	1957-1958
50	VAD1T; VAG1T; VAK1T VAD1U; VAG1U; VAK1U	T950G; U950G T950K; U950K T950D; U950D 57000-62934	1958-1959
50 Implematic irst Series	VAD1V; VAG1V; VAK1V VAD1W; VAG1W; VAK1W	V950G; W950G V950K; W950K V950D; W950D 63000-81126	1959-1961
50 Implematic econd Series	VAD3A; VAD3B	950/3A; 950/3B 400001-401489	1961-1962
50 Implematic irst Series	VAD2A; VAD2B VAG2A; VAG2B	A850D; B850D A850G; B850G 300001-306334	1960-1961
50 Implematic econd Series	VAD2C/36; VAD2D/36 VAD2M/36	850C; 850D 850DM 310001-317439	1961-1965
50 Implematic Narrow	VAD2CV/36; VAD2DV/36	850CV N390001-390469 850DV	1961-1965
80 Implematic irst Series	VAD2C/40; VAD2D/40	880C 350001-362382 880D	1961-1964

Description	Class	Serial Nos.	Production
880 Implematic Narrow	VAD2CV/40; VAD2DV/40	880CV N395001-395303 880DV	1961-1965
990 Implematic	VAD4/47A; VAD4/47B	990A 440001-480600 990B	1961-1965
880 Implematic Second Series	VAD2E; VAD2F	880E 521001-527521 880F	1964-1965
880 Implematic USA	VAD2UE; VAD2UF	880UE; 880UF	
990 Implematic USA	VAD4/47UA; VAD4/47UB		
Oliver 500	20C	101500-102148	1960-1964
Oliver 600	20A	100001-101500	1961-1964
770 Selectamatic	770A; 770B	580001-592375	1965-1970
780 Selectamatic	780/1	600001-611551	1967-1971
780 Selectamatic Narrow	780N	645001-645647	1969-1971
880 Selectamatic	880A; 880B	530001-563379	1965-1971
990 Selectamatic	990A; 990B	482001-505286	1965-1968
990 Selectamatic	990A; 990B	800001-831351	1968-1971
1200 Selectamatic	1200A	700001-718990	1967-1971
3800 Selectamatic	3800/1	650001-650522	1968-1971
4600 Selectamatic	4600/1	900001-900582	1968-1971
775 Synchromesh	775/1	594001-594235	1972-1976
885 Gasoline	885G	651001-651870	1972-1976
885 Synchromesh	885/1	620001-640365	1971-1976
885 Synchromesh Narrow	885/N	646001-647360	1971-1976
990 Synchromesh	990/1	850001-868541	1971-1976
995 Synchromesh	995/1	920001-936004	1971-1976
996 Synchromesh	995/6	980001-989042	1971-1976
1210 Synchromesh	1210/1	720001-731855	1971-1976
1212 Hydra Shift	1212/1	1000001-1005283	1971-1976
1410 Synchromesh 1412 Synchromesh	1410/1) 1412/1)	1050001-1051058	1974-1976
8 Series	885/1; 885/Q) 885/N; 885/G)	11000001-11021851	1976-1983
9 Series	990/1; 990/6) 995/1; 995/6) 995/Q; 990/Q) 995/6/Q)	11070001-11106274	1976-1980
12 Series	1210/1; 1210/Q) 1212/1; 1212/Q) 1210/8; 1210/8/Q) 1210/71;1210/78) 1212/71.	11150001-11167079	1976-1980
4 Series	1410/1; 1410/Q) 1412/1; 1412/Q)	11200001-11206445	1976-1980
880 UMU Diesel		91001-910048	1976-1980
4600 UMU		915001-915002	1969-1970
990 UMU Diesel		890001-890045	1969-1971